THE THAMES & SEVERN CANAL

a survey from historical photographs

by

D. J. Viner, B.A.,

Curator of the Corinium Museum

Cirencester

First edition – September 1975
Second impression – February 1977
Third impression – April 1981
Fourth impression – April 1982
Fifth impression – May 1989

Published by Hendon Publishing Co. Ltd., Hendon Mill, Nelson, Lancashire.

Printed by Peter Fretwell & Sons Ltd., Healey Works, Goulbourne Street, Keighley, West Yorkshire.

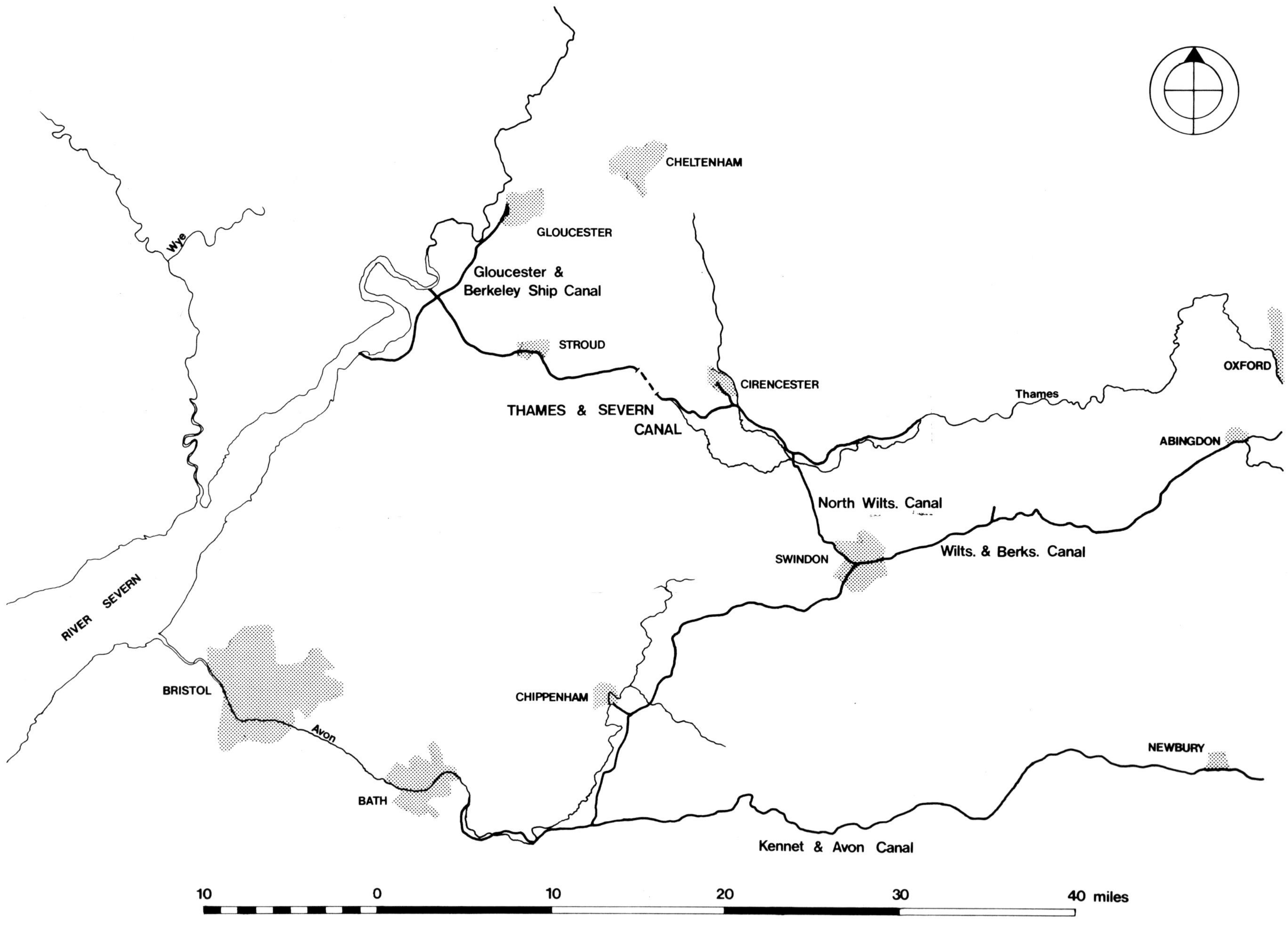
CHELTENHAM
GLOUCESTER
Wye
Gloucester &
Berkeley Ship Canal
STROUD
CIRENCESTER
THAMES & SEVERN
CANAL
Thames
OXFORD
ABINGDON
North Wilts. Canal
SWINDON
Wilts. & Berks. Canal
RIVER SEVERN
BRISTOL
Avon
CHIPPENHAM
BATH
NEWBURY
Kennet & Avon Canal
10
0
10
20
30
40 miles

THE THAMES & SEVERN CANAL...

commands an affection among enthusiasts which few other canals can match. Although the turbulent nature of its history and development is far from unusual, the T & S (as it is colloquially known) has always possessed character in abundance, and its route traverses one of England's most attractive regions, the Cotswold hills. The name alone conjures up the two great rivers, and the short length of nearly 29 miles belies the hazards of the Canal's construction and the difficulties of its navigation. Although now derelict throughout its length, the Canal is clearly visible in all but a few places, and much of it can be visited by car or on foot.

This survey is an attempt by the use of historical photographs to capture some of the atmosphere and character of the Thames & Severn Canal during and after its working life. It is in no sense a history of the Canal, or a detailed guide of its route, both of which lie beyond the scope of a study of this kind. Most of the photographs reproduced here are over fifty years old, and many have not been published before. Collected together, they offer a rare opportunity to bring the Canal into focus during the last years of the nineteenth and the early years of this present century.

The length of the Canal is followed in sequence, from west to east, the Severn to the Thames. From the junction of the Stroudwater Canal in Stroud, the T & S negotiated the narrow valley of the River Frome to Chalford, a landscape characterised by intensive industrial development and imposing mills dating from the period of Stroud's importance as a centre of the woollen cloth industry. From Chalford, the valley becomes rural and well justifies its name of 'Golden Valley'. The climb up to the summit level is negotiated through valley slopes heavily wooded on both sides, before Daneway is reached and the Canal disappears abruptly into Sapperton tunnel. Over 2½ miles long, the tunnel cuts through the Cotswold scarp and when built was the greatest civil engineering project yet attempted in Britain. From the eastern end of the summit level at Siddington, where there is a junction with the short arm to Cirencester, the Canal falls gently (and circuitously) through the open farmland of the Gloucestershire/Wiltshire border for 13½ miles, to reach the Thames at Inglesham, near Lechlade.

The relation of the T & S with the rivers Severn and Thames and with other routes on the canal system is shown opposite; for a detailed plan of the Canal showing locks and bridges, see the centre pages.

The two great rivers of southern Britain are the Severn and the Thames. In modern times, it could be argued that the former has been overshadowed by the latter, but this has not always been the case. In the late seventeenth century one third of the total of 300,000 tons carried on Britain's rivers was transported to the several important towns on the Severn. Much of the agricultural and industrial goods needed to supply the city of London was produced near and carried upon the Severn; and Bristol was the second port of the kingdom.

It was logical and economic good sense to seek a connecting route between the river Severn and its link to London, the Thames. The barrier was the Cotswold scarp, broken up into narrow valleys but nowhere allowing an easy route through into the headwaters of the Thames. Of the alternative routes, the more southerly, along the Avon valley, offered the most direct route between Bristol and London (and was subsequently used for the Kennet and Avon Canal), but the central route along the Stroudwater or Frome—the 'Golden Valley' route—was eventually chosen for the Thames and Severn Canal.

Proposals to link the two rivers date back at least to the early seventeenth century. There was considerable interest at this time in rendering rivers more navigable, and no less than five bills were introduced in the House of Commons in the 1660's, all without success. But interest continued to build up in the following century, during which the woollen industry of the Stroud valleys reached the height of its prosperity. In 1730 an act was passed for the construction of a 'new cut' to Stroud, and although it was strongly opposed by mill-owners fearful for their water supplies, the canal was eventually begun 45 years later and the Stroudwater Navigation opened in July 1779.

A summary of the development and construction of the Thames & Severn Canal can be given in gazetteer form:
17th September 1781: a public meeting was held in the Kings Head Hotel, Cirencester, when a committee was appointed and a subscription list opened to build the Thames & Severn Canal. The Chairman was Edward Loveden, owner of Buscot Park near Lechlade, a leading member of the Thames Commissioners (responsible for the upkeep of the river above Staines) and a powerful supporter of new canals.

22nd December 1781: the proprietors met Robert Whitworth at Cirencester and commissioned him to survey a route. Whitworth had been an assistant of James Brindley, the great canal builder, and he was already a highly regarded surveyor for several canal schemes. Whitworth's proposed route was at an estimated cost of £128,000.

1. Recovery . . . narrow boat *The Trial* shown between the two Walbridge locks and about to leave the Thames & Severn for the Stroudwater Navigation.

The date is 11th April 1899 and *The Trial* is returning to Stonehouse from London with a cargo of wool for Charles Hooper & Sons, millowners at Eastington. Her round journey took three weeks and resulted from the reopening of the canal by the Thames & Severn Canal Trust.

2. Dereliction . . . the summit level shown very dry in 1896.

From the outset, the Canal's engineer Robert Whitworth was aware of the 'bad Rocky Ground' on the summit level, a parched upland area where water disappeared all too easily into the porous limestone. Shortage of water was a permanent problem.

As a result, the various restoration attempts in the years following the date of this photograph were short-lived.

17th April 1783: the bill to 'make and maintain a navigable canal' from the Thames at Lechlade to join the Stroudwater Canal at Stroud was passed through Parliament.

June 1783: construction work began at the Stroud end, moving steadily eastwards. This enabled materials required for the construction work to be transported close to the sites; stone, clay and timber were readily available locally and brickyards and limekilns were established when required alongside the works. The first vessel entered the T & S at Walbridge in January 1785, by which date the canal was probably completed as far as Chalford. The steady climb to the summit level involved the construction of a large number of locks (fifteen from Chalford to the top) but the summit lock and basin at Daneway were open by the summer of 1786.

1784: early in the year work began on the digging of Sapperton tunnel, 'the most ambitious engineering feat of its day' according to Household. Certainly, it was the longest to be built and even today remains as the third longest canal tunnel ever constructed at 3,817 yards in length, 15 feet in width, and 15 feet in height. Whitworth's estimate for construction costs was £36,575 but this figure was considerably exceeded. The tunnel took five years to complete, as a result of incompetence and bad management on the part of the major contractor, Charles Jones. Progress was erratic until his removal, when five groups of contractors began a concerted effort towards completion.

19th July, 1788: much of the structure of the tunnel had been completed when King George III and his family paid a visit to the work, inspecting the classical eastern portal as guests of Earl Bathurst. The King was impressed at the achievement which the tunnel represented. In April of the following year the first boat passed through the tunnel.

Meanwhile work had been progressing with relatively little difficulty on the remainder of the canal's length. The summit level from the entrance of the tunnel to Siddington, together with the Cirencester branch had been completed by 1787 and the long eastern section to within three miles of the Thames by April 1789.

19th November, 1789: the first boat passed through the canal into the Thames, on a day of celebration in nearby Lechlade, with dinners, a bonfire and a ball and a twelve cannon salute from Buscot Park. The construction of the Thames & Severn had taken only 6½ years from the passing of the Act.

Building the Canal had been an expensive business and the Company remained in debt until a re-organisation of its finances enabled a first dividend to be paid in 1810. The Canal had been envisaged as an east-west link route but the passage of goods from Severn to Thames and vice-versa remained fraught with difficulties; navigation of the Severn was difficult and the administration of the upper Thames was in a state of disorganisation. New canal schemes overcame most of these problems. The opening of the Kennet & Avon and the Wilts & Berks Canals in 1810 provided a rival and alternative route between the two great rivers; perhaps its greatest advantage was the evasion of the notorious upper reaches of the Thames by joining the river at Abingdon. The construction of the North Wilts Canal in 1819 linked the T & S to this route and allowed its trade too to enjoy a less difficult route to London. The final link was the construction of the Gloucester & Berkeley Canal, opened in 1827, which joined the Stroudwater on the level at Saul Junction and enabled the difficult stretch of the Severn between Sharpness and Gloucester to be avoided.

The finances of the Thames & Severn Canal Company show that its most prosperous period was the third and fourth decades of the nineteenth century, reaching a peak in 1840-42. In 1841 an income of £11,000 was achieved and the maximum tonnage carried on the canal recorded at 89,271 tons. However, the boom of these years partially resulted from the development of the canal system's worst enemy, the railway.

In 1841 the GWR line from Swindon to Cirencester was opened and completed to Gloucester in May 1845. Henceforth the railway was adversely and dramatically to affect trade on the Canal—receipts had dropped to £3,000 in 1855.

10 years later came the first attempt to turn the canal into a railway, using the length between Stroud and Cerney Wick and including the tunnel at Sapperton. Ever-fearful of competition in the heart of its area, the GWR attempted to rent the canal, but the conversion plan failed to materialise, and similarly a subsequent scheme in 1882.

But the influence of the GWR was certainly not at an end and the canal passed into the railway company's ownership, to be followed by the closure of the 26 miles from Chalford to Lechlade. Local protestations followed, justifiably concerned at the GWR's motivations, and the resulting Act of 1895 transferred control of the T & S to a Trust, consisting of representatives of all local authorities and interested canal companies. With £19,000 to spend on restoration, the Trust successfully restored navigation in 1895 but by 1901 the continuing problem of excessive leakage of water caused further abandonment.

Already involved, Gloucestershire County Council assumed managerial responsibility and further expenditure was authorised in a restoration programme from 1901-7. Certain sections and locks were completely relined but there was little traffic to create sufficient demand, and on May 11th, 1911, the last reported boat, with a load of stone, passed through the tunnel. The County Council's expenditure in that decade had amounted very nearly to £30,000. The rest is quickly told; the Canal was finally abandoned in two stages, eastwards from Whitehall Bridge to Lechlade in 1927 and completely on the 9th June 1933.

This study perhaps accentuates one aspect of the T & S more than any other— the interesting range of buildings alongside the towpath. In terms of architectural history, the Canal could be seen as a classic example of its kind, showing the development through stages of a functional group of buildings.

It is doubtful whether an architect—in the modern sense of the word—was involved in the Canal's buildings, and this possibly makes them even more interesting. Many, indeed, so closely resemble the standard Cotswold building types around them that they are virtually indistinguishable except by their canal-side position, often isolated in the fields. There is nothing remarkable, for example, about the little cottage by the basin at Daneway (no. 25), or its twin at the entrance to the tunnel (no. 26).

The construction of the Canal involved an enormous amount of working in stone, and gangs of masons were everywhere at work. For them, the erection of buildings was an integral part of their daily toil.

But the T & S is important because it is an example of the deliberate stimulus given to this strong local tradition of the stone-mason's craft. At specific and well-chosen points, the opportunity was taken to make emphasis of the structure being built, with a striking design—hence the portals of Sapperton tunnel. The western end is gothic (no. 27) and is an austere beginning to what must have been a frightening journey through the tunnel.

The eastern entrance is classical (see drawing on page 24) and perhaps the more impressive. There is little doubt that the influence of the landowner Earl Bathurst played some role in the choice of design here. Masons familiar with work on the many large, classical houses in the surrounding Cotswold towns, and especially Cirencester, would have experienced little difficulty at the tunnel.

Equally remarkable are the round houses, five altogether (e.g. no. 16). In all other respects, these supplied the usual functions as cottages for lengthmen, but they were architecturally pretentious in shape. One can imagine that round houses presented a challenge to the masons, but this was overcome and all five survive.

The other group worthy of mention are the three wharfhouses, at Cirencester (nos. 37 and 41), Cricklade (nos. 50 and 51) and Kempsford. They are obviously functional buildings in shape, with large warehouse space at sides and rear, but this was finished off with a classical facade, the sole purpose of which was to impress.

A journey along the Thames & Severn was full of surprises, with changes in scenery and landscape almost dramatic in places, a long and mysterious tunnel, and at regular intervals fascinating peculiarities in the design of many of the Canal's buildings.

3. In the narrow valley of the River Frome road, canal and railway are never far apart. In this 1955 view, the river is close at hand to the right, and the railway crosses overhead. The bridge and viaduct carries the main ex-G.W.R. Gloucester-Stroud-Swindon line; in the background, the second bridge allowed access to the ex-Midland Railway goods yards.

'Industrial' valley?

Two views within $1\frac{1}{2}$ miles of the centre of Stroud. The mills for which the area is famous are scattered along the valleys with Stroud as the nucleus; hence stretches of open countryside and farmland.

4. Looking east to Stanton's Bridge between Bowbridge lock and Griffin's Lock, which can just be seen under the bridge-arch. Basically an accommodation bridge, Stanton's enjoys a central position on the first restored stretch of the canal (see no. 55). Seen in 1955.

5. Griffin's Lock in 1896, with Stanton's Bridge in the background. Just visible on the right is the overflow channel for the lock.

The mill after which the lock is named survives close by—a nineteenth century building in use for many years as a timber mill.

6. There are comparatively few pictures of working activity on the Thames & Severn, and the one opposite is a remarkable survivor. It shows the nb *Alert* moored by the stern to the top gates of the lock variously known as Gough's Orchard, Dalloway's and Lewis' Lock, after the respective owners of the mill alongside.

Two Messrs. Davis, both of Chalford but no relation to each other, are seen unloading coal into Brimscombe Mills, then in use by P. C. Evans & Sons, in the years before the First World War. A. M. Pearce was a well-known carrier in the Stroud valley and the *Alert* was one of his boats.

Two views of an obliterated section . . .

7. (top right) shows Brimscombe Bridge, looking west, and a typical local subject for an early twentieth century postcard. The family group walking along the towpath pause for the camera, probably upon request, but without doubt the figure on the bridge simply appeared at the appropriate moment! The brick-built bridge has gone and the canal filled in, but the Ship Inn remains.

8. (bottom right) an unhappy result to a well-meaning attempt to record the approach to Brimscombe Port from the west. Although the canal has been filled in, its line survives to show the characteristic sweeping approach to the Port buildings in the background.

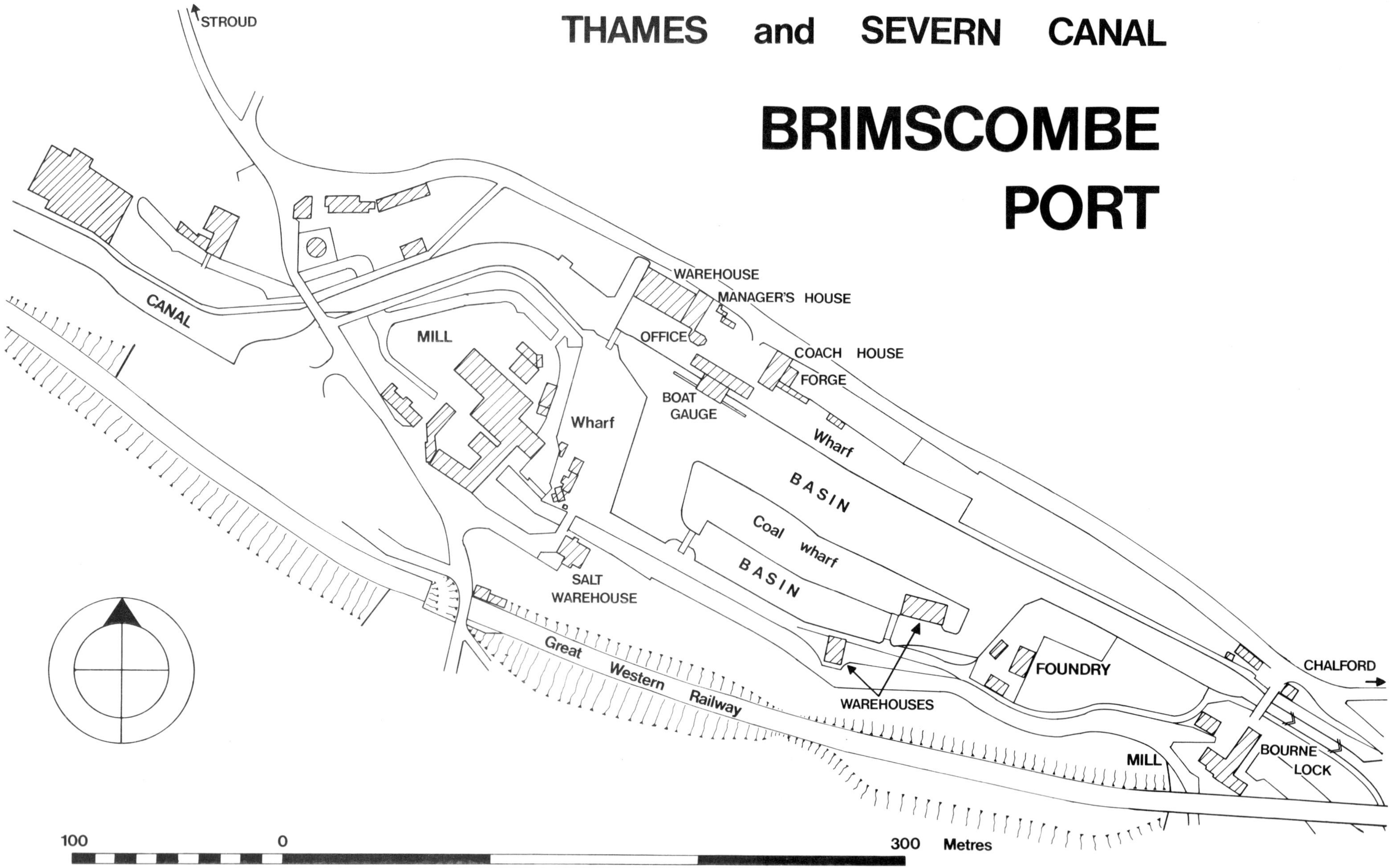

THAMES and SEVERN CANAL
BRIMSCOMBE
PORT
STROUD
CANAL
MILL
WAREHOUSE
MANAGER'S HOUSE
OFFICE
COACH HOUSE
FORGE
BOAT
GAUGE
Wharf
Wharf
BASIN
Coal wharf
BASIN
SALT
WAREHOUSE
Great Western Railway
WAREHOUSES
FOUNDRY
CHALFORD
MILL
BOURNE
LOCK
100
0
300
Metres

The importance of Brimscombe Port in the life of the Canal is obvious from the plan of its buildings (opposite) and the scene of activity in 1826 (see front cover).

All the major buildings were sited here, with ample provision of warehouses and wharves, including the famous coal wharf, sited on an island in the centre of the basin in an attempt to alleviate pilfering.

9. Is a view of the most famous Port building, the great warehouse on the northern side of the basin, with the agent's house and offices at one end. The 1826 view shows the building in its original state, but in 1911 it was converted for the use of Brimscombe Polytechnic, with the insertion of ever-increasing window-space.

10. Not the least interesting feature of the Port was the boat gauge, erected in 1845. It was literally a weighing machine, consisting of a cast-iron cradle in a massive superstructure, and capable of recording accurately the weight of the boat plus cargo. As a deterrent to toll-evasion, it was very successful and justified its cost of over £1,000. The photograph was taken during dismantling in the summer of 1937; the original model for the machine is in Gloucester Folk Museum.

11. Another postcard subject, and an invaluable record of both canal and railway. The scene is Beale's Lock just below Brimscombe G.W.R. station, and the date sometime in the 1920's. The push'n'pull service between Stroud and Chalford is seen approaching Brimscombe on the return journey to Stroud, with the tiny locomotive shed on the left.

Evidently a hot summer afternoon, and with only minor alterations the scene was much the same up to forty years after the date of this photograph.

12. The same subject in 1957, this time looking west towards the station from the towpath. The top gates on the lock have been replaced by a concrete barrier, and hence the good level of water above the lock.

Barely distinguishable through the trees, an ex-G.W.R. pannier tank climbs through the station with a coal train.
Another hot summer's day?

Approaching Chalford, the valley grows narrower, and buildings cluster even more closely to the canal-side.

13. Looking east with Chalford Chapel in the middle distance. Several locks follow each other in quick succession here; the bottom gates of Grist Mill Lock show beneath the first bridge, and those of Ballinger's Lock beneath the next bridge up.

This photograph comes from the collection of Frith & Co. of Reigate and dates from about 1910.

14. A closer view of Ballinger's Lock, again looking east, and of much the same date. The lock-gates are in excellent order following their replacement in 1902. Both the lock and the group of canal-side buildings in the background have disappeared and garages now stand on the site of the lock. The distinctive conical roof of Chalford round house is just visible on the right, below the chimney.

15. One of the Canal's most attractive spots. April 1937 at Chalford, with the roundhouse on the right undergoing external refurbishing. It clearly remains occupied! On the left, the bridge across the Canal just below Chapel Lock. A tranquil scene, rarely disturbed by passing boats, but well-kept nevertheless.

16. There were five of these distinctive roundhouses on the Thames & Severn. Two—this one at Chalford and a second at Cerney Wick (see Nos 46-7)—boasted the attractive external conical roof. They served as cottages for lengthmen, or maintenance men, and are evenly distributed along the line of the canal.
Living accommodation was on the first floor and sleeping accommodation on the second. The ground floor acted as a semi-basement and had easy access to the road alongside.

Behind the roundhouse was The Company's Arms Inn, a fascinating building with a long history. Its earliest features date from the fifteenth century, and it was still in use as an inn when this photograph was taken in 1955.

17. A later view of the same scene as opposite. By 1955, the canal banks had become overgrown and the canal itself clogged with weed.

In 1966, the canal was culverted and infilled and the Chapel bridge demolished. However, the area around the roundhouse was landscaped as an amenity and the canal still holds water here.

18. A rarely-photographed lock, Chalford Chapel Lock, seen here at Easter 1933, the year in which the Canal was finally abandoned. The shortage of water, so much a problem elsewhere on the route, does not seem to present a problem here.

The bridge was built on a skew to the canal, and was of red-brick, with the exception of a remodelled section on the right-hand side in local stone. The balance-beam betrays the age of the lock-gates.
This section was culverted in 1966.

19. A turn of the century view looking west towards Chalford Chapel in the left background. This curving section has now been culverted and the site on the left-hand side is now part of the Chalford industrial estate.

The horse-bus provides an interesting incidental feature; the line of the modern road preserves the curving alignment of the canal.

20. Bell Lock, leaving Chalford for the Golden Valley. Originally a Frith & Co. postcard, the date is about 1910. Both lock and inn have gone, but the Red Lion Inn in the background survives. At the time of this photograph it belonged to the Nailsworth Brewery.

East of Chalford, and still steadily climbing the valley.

21. Looking west down to Red Lion Lock, a little way above the inn after which it was named.

In the foreground, Sevill's or Clayfield Mill, finally demolished in 1964.

Taken from an early postcard, the date of which is probably very early this century.

22. Golden Valley Lock, the next above Red Lion Lock, with the Valley Inn on the left. The railway is apparent again in the left background. About 1910.

After the failure of the Thames & Severn Canal Trust to restore the canal to navigation at the end of last century, the responsibility passed to Gloucestershire County Council and a second restoration attempt was embarked upon in the first decade of this century. The total declared cost was nearly £30,000, but again the problems proved insurmountable and navigation was only restored for short periods at a time.

23. Puck Mill Upper Lock, seen during this period of restoration, but before any work had actually begun on the gates of the lock. On the left, an inn, The Oak, typical of isolated canal-side pubs relying almost totally upon passing boatmen for their trade. The building has long since ceased to be an inn.

24. Puck Mill Lower Lock. The date is 1907 and Glos. C.C. workmen are busily re-puddling the bed of the canal and re-building the approaches to the lock chamber. This extensive work was not part of the original County Council programme but was made necessary by the fact that the pound between the two locks 'emptied itself about every four hours', according to one local newspaper report.

Looking down the valley, Baker's Mill can be seen in the middle distance.

25. At Daneway (left), the canal reached the summit level, 28 locks and just over seven miles from Stroud. The tiny settlement was almost totally canal-orientated and centred upon the inn. Originally called the Bricklayers Arms, the Daneway Inn sprang up with the arrival of the construction workers engaged for five years upon the cutting of the tunnel close by. The basin at Daneway was the upper terminus of the navigation for several years until the tunnel was completed.

This turn of the century postcard view shows the basin looking east; in the centre Daneway bridge conceals the top lock just beyond, but the inn is clearly to be seen on the left. Although the basin is overgrown, all the buildings and the bridge survive, and the Daneway Inn is a deservedly renowned local hostelry, still managing to avoid the hammer of modernisation.

26 & 27. (top and bottom right respectively). The western or 'Daneway' entrance to the Sapperton tunnel. Both portals form famous and much-loved landmarks on the Thames & Severn, although both are in deteriorating condition. No. 26 shows the two maintenance cottages on the right in a postcard view dating from about the turn of the century. The left-hand cottage remained in occupation until recently, despite being some way from the road. No. 27 probably dates from the Glos. C.C. restoration period early this century, and shows clearly the bold design which characterises the architectural pretensions of the Canal.

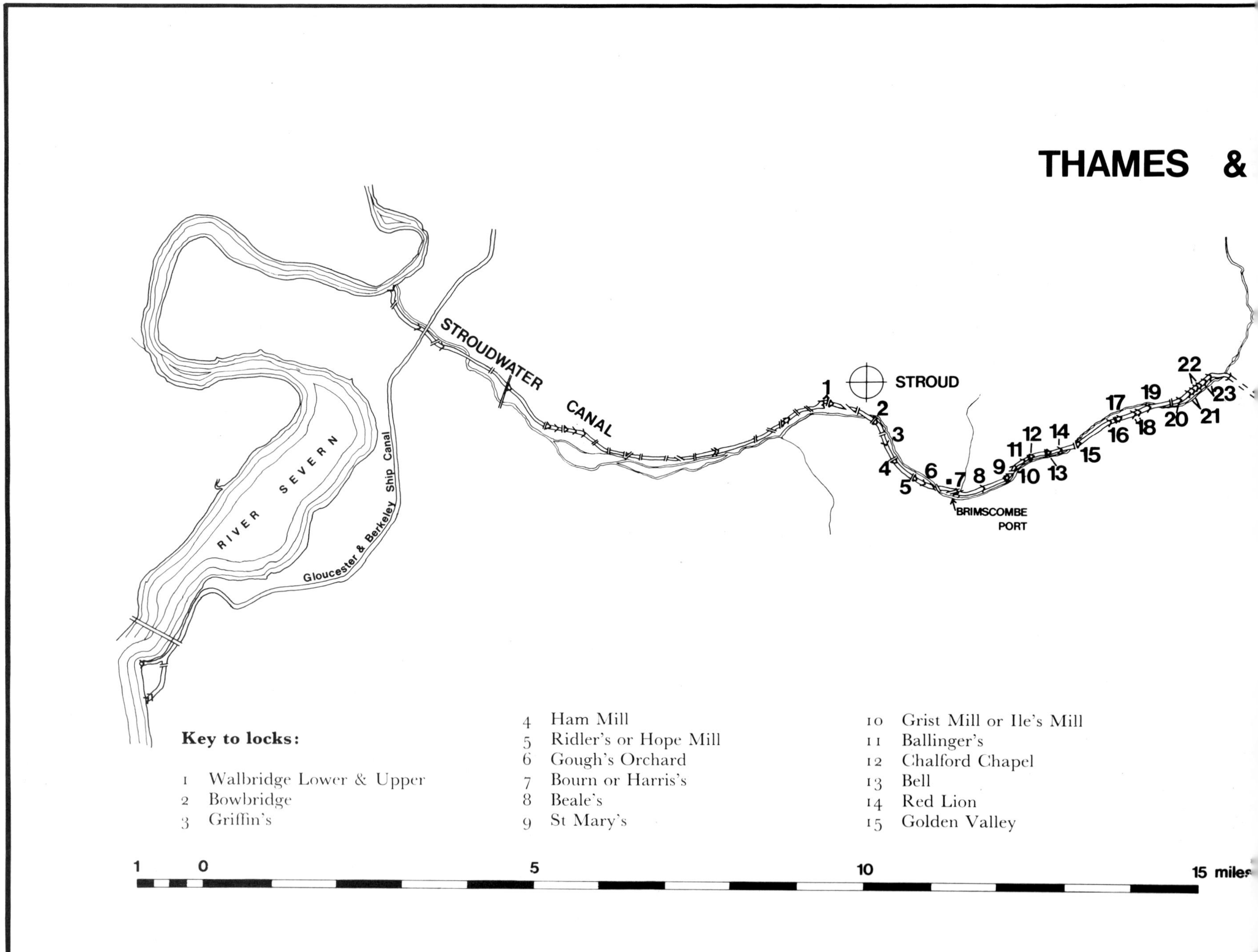
THAMES &
STROUDWATER CANAL
STROUD
BRIMSCOMBE PORT
RIVER SEVERN
Gloucester & Berkeley Ship Canal
1
2
3
4
5
6
7
8
9
10
11
12
13
14
15
16
17
18
19
20
21
22
23
Key to locks:
1 Walbridge Lower & Upper
2 Bowbridge
3 Griffin's
4 Ham Mill
5 Ridler's or Hope Mill
6 Gough's Orchard
7 Bourn or Harris's
8 Beale's
9 St Mary's
10 Grist Mill or Ile's Mill
11 Ballinger's
12 Chalford Chapel
13 Bell
14 Red Lion
15 Golden Valley
1 0 5 10 15 miles

SEVERN CANAL

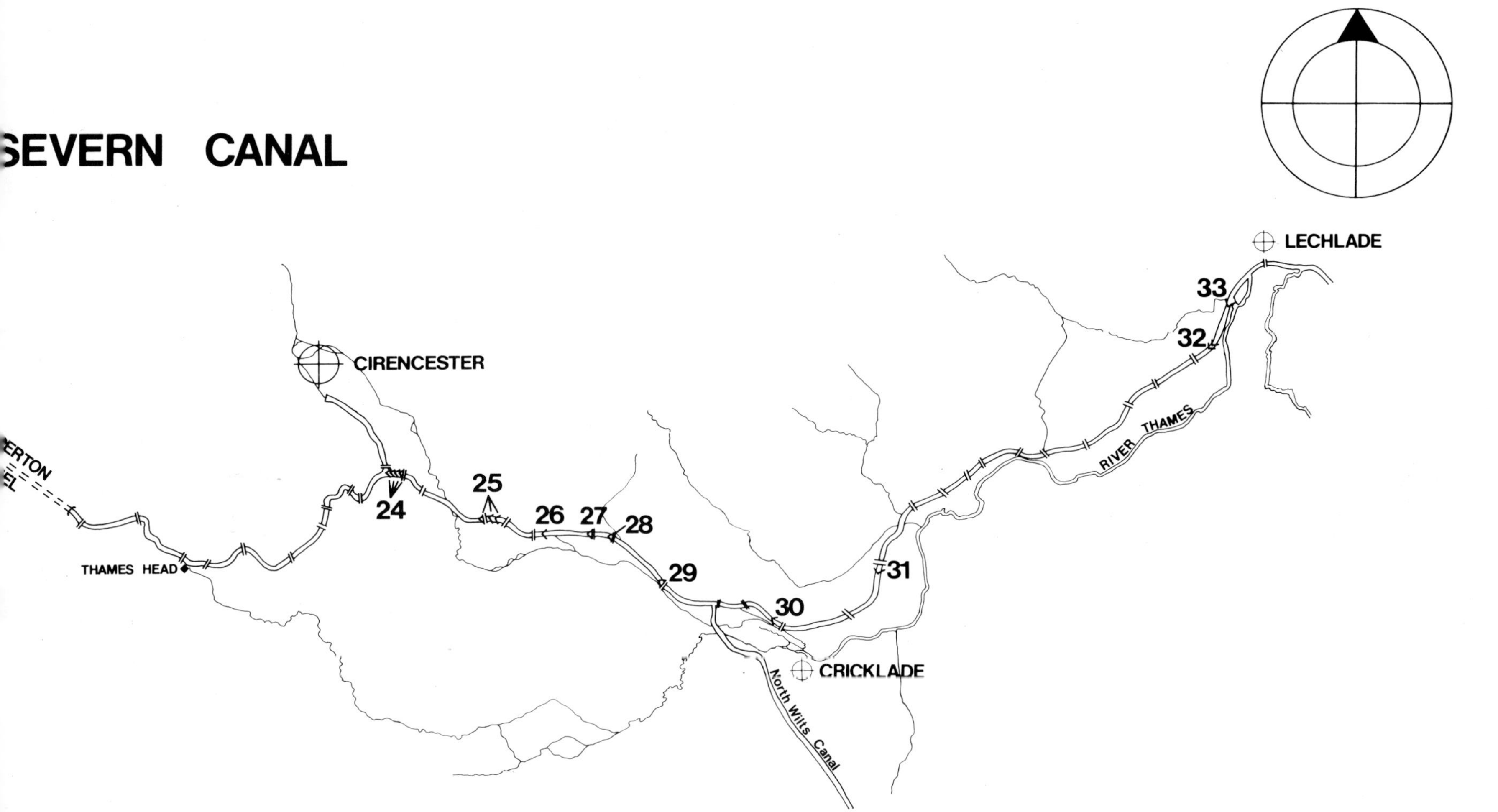

Baker's Mill Lower
Baker's Mill Upper
Puck Mill Lower & Upper
Whitehall Lower
Whitehall Upper
Bathurst's Meadow Lower & Upper

22 Sickeridge Wood Middle & Upper
23 Daneway Lower & Upper
24 Siddington 1 to 4
25 South Cerney 1 to 3
26 Boxwell Spring
27 Wilmoreway Upper
28 Wilmoreway Lower
29 Cerney Wick
30 Latton
31 Eisey
32 Dudgrove Double
33 Inglesham

JB '74

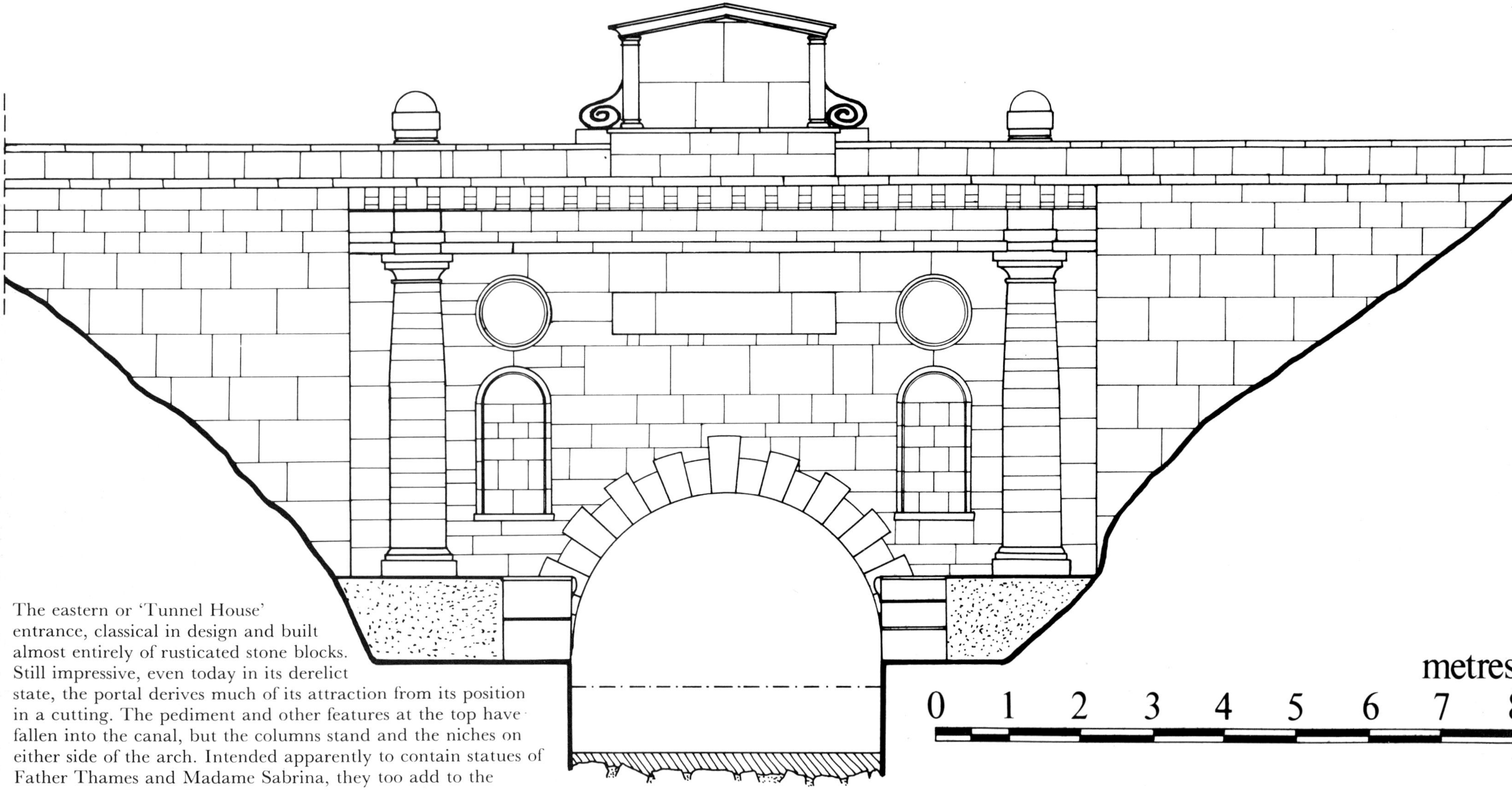

The eastern or 'Tunnel House' entrance, classical in design and built almost entirely of rusticated stone blocks. Still impressive, even today in its derelict state, the portal derives much of its attraction from its position in a cutting. The pediment and other features at the top have fallen into the canal, but the columns stand and the niches on either side of the arch. Intended apparently to contain statues of Father Thames and Madame Sabrina, they too add to the symmetry of the whole.

SAPPERTON TUNNEL: EASTERN PORTAL.

28. The Tunnel House Inn, just above the eastern portal of the tunnel. It is seen here in April 1947, very much in the original condition in which it survived until a disastrous fire in 1952. Built by Lord Bathurst as the New Inn, it served a similar purpose to its counterpart at Daneway as a lodging house for the tunnel's construction workers, and for this the upper floors were divided up as dormitory space, and the single ground-floor tap-room made large enough to serve as a communal dining room.

The Inn was rebuilt from the fire's ruin and has become a popular haunt for Cotswold visitors and canal-enthusiasts alike.

29. Tarlton road bridge, as rebuilt in 1823. Relatively few of the Canal's bridges possess this dignified design, stone-built to complement the surroundings. Beyond the bridge was the King's Reach, a stone-cut channel extending to the tunnel mouth, and impressive in its proportions. The name resulted from a visit to the works in July 1788 by King George III and his family.

This section was re-puddled by Glos. C.C. and the photograph probably dates from this period.

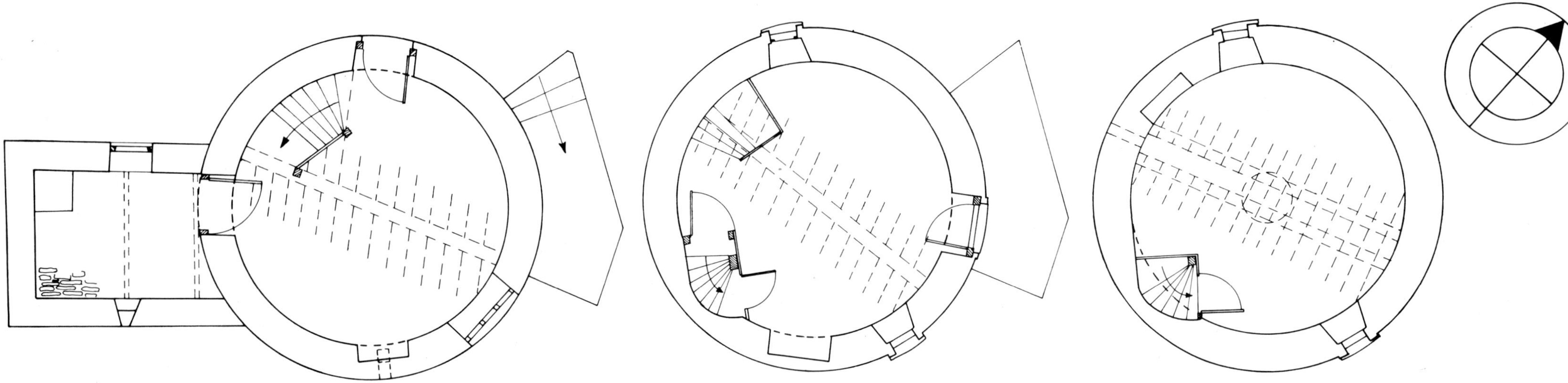

LENGTHMAN'S COTTAGE,

Thames & Severn Canal,

Coates, near Cirencester.

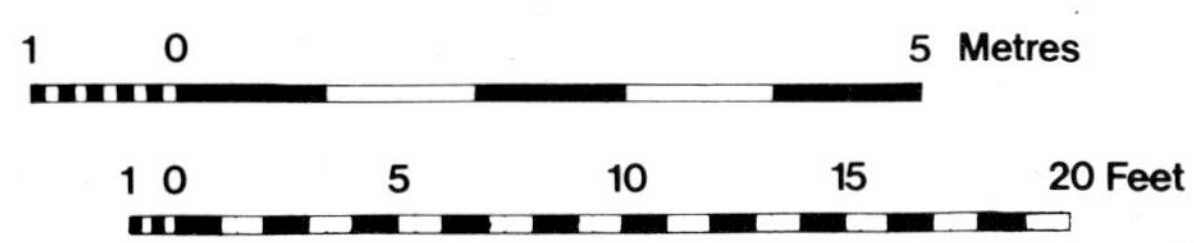

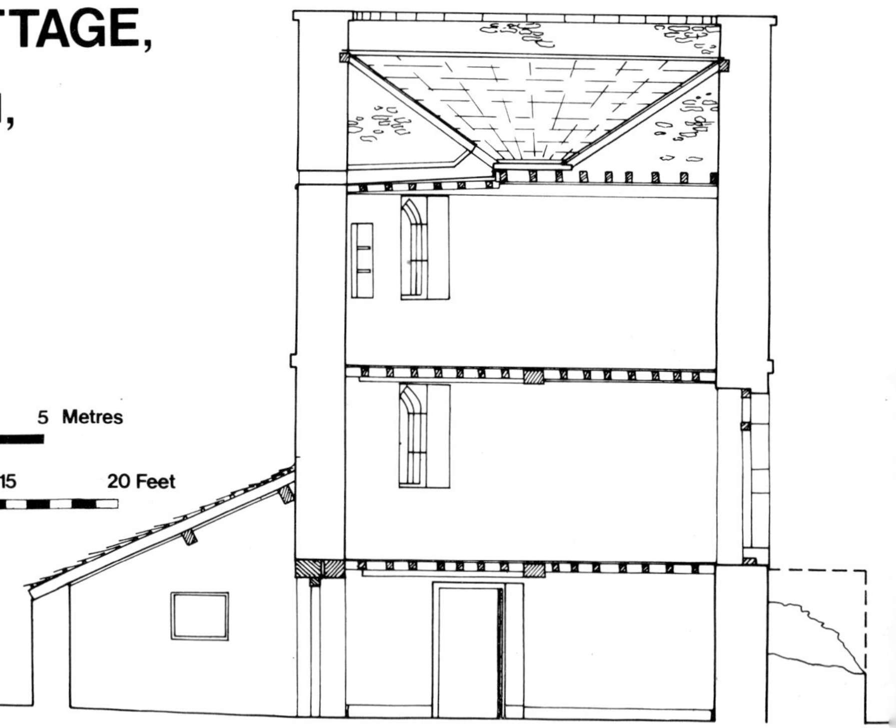

RW,J

30. A fascinating building, the roundhouse at Coates, seen here whilst still in use as a cottage in August 1949. The plan opposite shows clearly the inverted conical roof design, intended as a large water trough. All three roundhouses with this feature (see Nos. 53-4) are in isolated positions, and doubtless benefited from this seemingly disturbing feature!

Note the conversion of the ground-floor area to living accommodation and the insertion of a window. The scullery added on to the building upsets the symmetry but can be excused as a (relatively) modern refinement.

31. An attractive view of Coates roundhouse in 1938. Emerging from the King's Reach, the Canal suddenly finds itself on the open Cotswolds, and this next section is Whitworth's 'bad Rocky Ground', so fraught with difficulties of water retention. In this view, the canal bed is almost totally dry; note the position for the stop-planks close to the roundhouse. In the distance, the bridge carries the Gloucester-Stroud-Swindon railway line.

32. The subject of the engines at the Thames Head pumping station would occupy many hours of fascinating study.

This is perhaps the most well-known view, dated to 1828 and showing the second installation on the site, containing a Boulton & Watt single acting beam engine in use from 1792 to 1854.

In the left background, the spire of Kemble church can just be seen.

33. Smerrill aqueduct, a single-arched masonry structure crossing the main Kemble-Cirencester road, and now removed for road-widening. Notoriously leaky, the aqueduct was of nuisance value to the locals passing underneath, and doubtless this farm-worker (complete with hay-fork and 'yorked-up' trousers) would have been specific on the subject.

From an old postcard dated c.1914 and looking towards Kemble.

Three views of the G.C.C. restoration programme earlier this century, at Furzenleaze, near Siddington.

34. (top right) This section is known as Blue House Reach, and the building in question can be seen just beyond the bridge. The workmen are cutting clay to be used in puddling.

35. (bottom right) These views are inter-related, and this shows the same area, with the group of men in the background of no. 34 now in the foreground.

The crushed clay is transported in skips along the track, which crosses the Canal under the bridge, to . . . .

36. (facing page) Again from the bridge, but now looking north. Repuddling is in progress, with the canal dammed on either side of the workings.

Blue House is on the right, and the Cirencester-Kemble GWR branch line crosses the photograph in the middle distance. September 1904. All three photographs suggest that they belong to an official record compiled on the restoration programme.

AGENT'S HOUSE,

Thames & Severn Canal,

Quern's Hill, Cirencester.

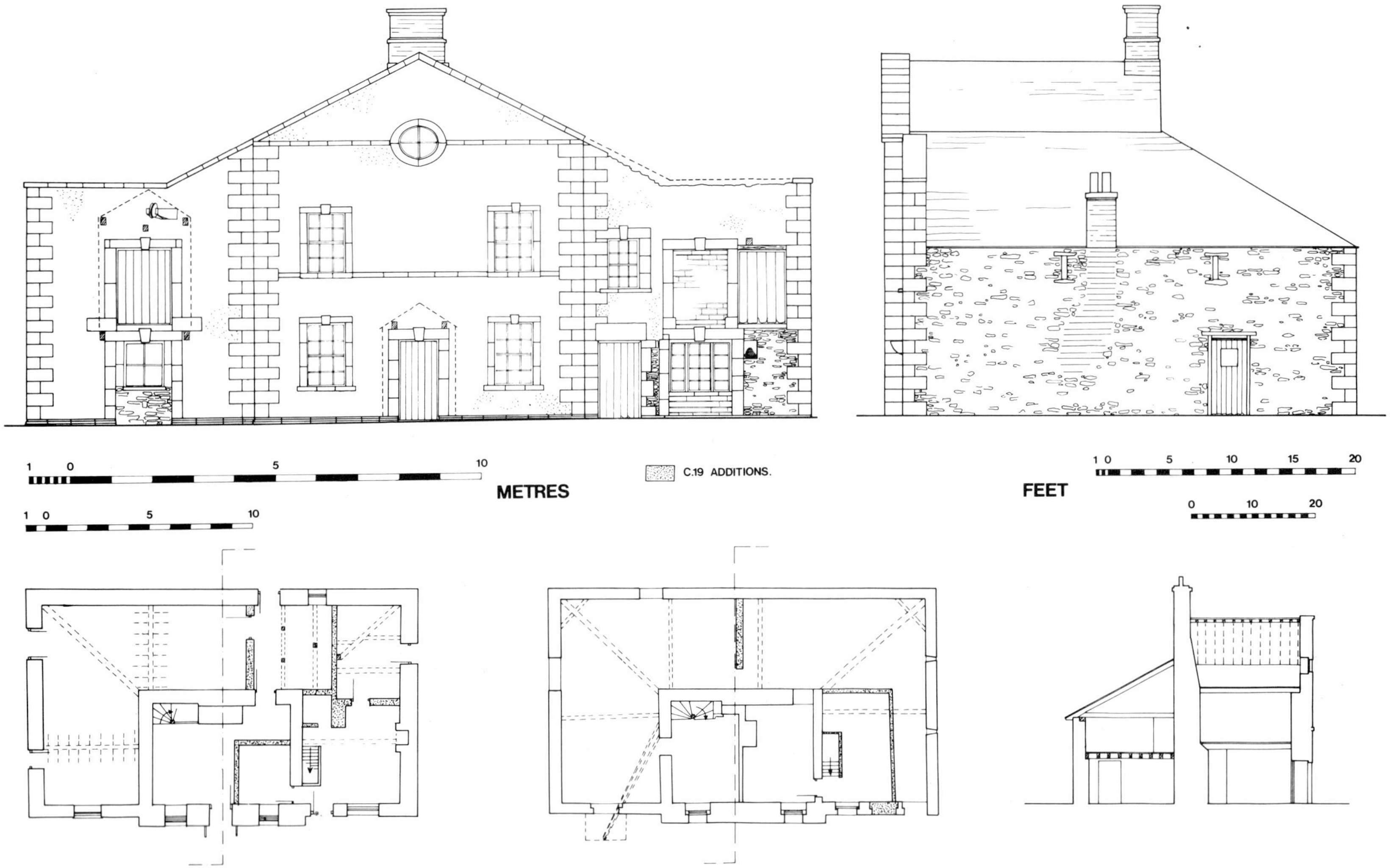

37. Cirencester wharf, showing the classical facade of the wharf agent's house-cum-warehouse at the rear. By 1904, when this picture was taken, the building had been much altered and the living space extended into the storage area on one side (see opposite) by the insertion of extra windows.

The occasion is a celebrated one: the arrival of the nb Staunch with a load of coal, the first vessel to reach the wharf for 16 years. This was a direct result of the restoration programme, but this scene remained exceptional.

Frank Gegg & Co. was a local Cirencester canal business, operating from the wharf for several years.

38. The previous picture shows probably all the company's waggons being loaded, one of which is seen here on the weighbridge at the wharf.

39. Both photographs show Frank Gegg's horse "Joey" and the figure here is almost certainly Gegg himself. Long after the wharf had ceased commercial use, he poses with "Joey" for the camera, and perhaps both are in retirement. Probably dated to the early 1920's, the wharfhouse is still used as a cottage.

40 & 41. These photographs were taken in September 1974, at the very end of the building's life. The wharfhouse was due for demolition during 1975.

The structure of the building can plainly be seen in its dereliction. The facade is random stonework, with very few blocks, and only the quoins offer solidity. For its size, these warehouses offered a great deal of storage space, and on the first-floor this extended right round the building from one wing to the other.

The conversion of the east wing into living accommodation is apparent, the original opening is blocked up and a new one built. A window and door have been inserted. Note the stucco finish to the facade.

42. From Siddington, the Canal began the long descent to the Thames. This view dates from the first decade of this century and shows the flight of locks at Siddington, looking up from No. 3 towards No. 2 lock. Beyond the bridge in the background is a T-junction between the main line and the Cirencester arm. This picture shows a healthy-looking canal, well-maintained and well-watered. It probably post-dates the restoration work for this reason. Now derelict and overgrown, the locks remain and the bridge too—the houses in the background form the community of Upper Siddington.

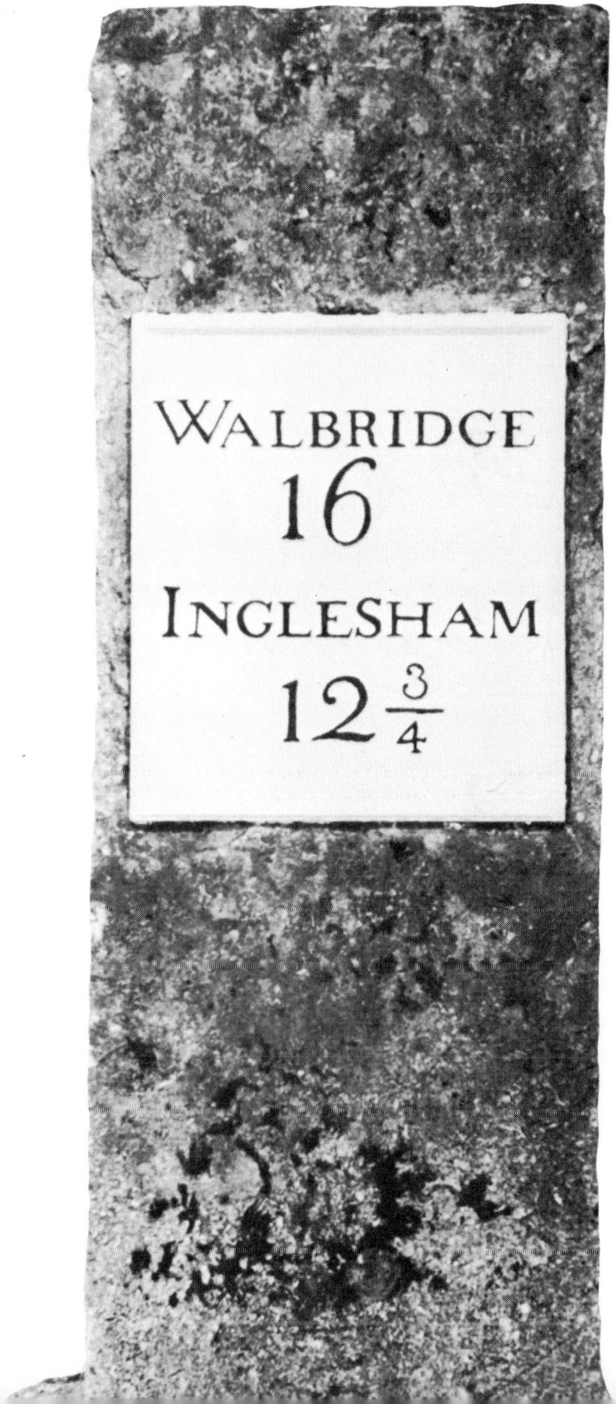

43. Milestones are features perhaps more normally associated with roads rather than canals, but their use was widespread on the latter and for a similar reason—tolls were levied on distances as well as upon loads carried.

There was a stone every half-mile on the Thames & Severn, measured in the mile and half-mile from the Stroud basin at Walbridge. The result was that the measured distance from the Thames end, at Inglesham, was at $\frac{1}{4}$ and $\frac{3}{4}$.

This example preserved in Corinium Museum at Cirencester comes from Siddington and was one of the rare surviving examples of a stone with its plate intact; the cast iron plate is frequently a target for undiscriminating souvenir hunters, with the result that few originals survive complete in situ.

The stone measures 1.25 m in total, of which 0.90 m was normally above ground. The top of the sunken portion is apparent where the stone begins to swell out at the foot of the photograph. Not all stones were squared off—several examples have attractive round heads.

At South Cerney, a flight of three locks continued the descent begun at Siddington.

44. Shows the top lock and lock cottage, taken from a postcard and dated to 1910-12. The lock-keeper, Mr. Tuck and his wife and family stand in front of the cottage, which has an interesting bay-window arrangement. A cellar gave storage space for goods.

45. Three men at leisure on the canal, entering the bottom of the three locks at South Cerney. Middle lock is in the background and the lock-cottage is just visible through the trees.

An early view, dated to c.1875 and often regarded as an early example of pleasure boating on canals.

46 & 47. Cerney Wick lock and round-house. The second of the upright conical types (see no. 16) this round-house is well preserved as a cottage.

Both views belong to the restoration period of early this century, note the new balance beams on the top gates. Probably 1900-5. One of the two gentlemen in the lower picture was undoubtedly the lock-keeper.

Two early and indistinct views of a section of the canal now fast disappearing. Both date from 1896.

48. Shows the junction with the North Wilts Canal at Latton Junction. From here, it was a short nine-mile journey to the Wilts & Berks Canal in Swindon and thence to the Thames at Abingdon. The stop-gate can just be seen and, on the left, the junction cottage.

49. Eisey, where a lonely cottage stands at a small wharf. Still surviving, the cottage would pass un-noticed in the Cotswolds were it not for its canalside position.

The canal now travels through the dairy lands of the upper Thames valley.

50 & 51. Cricklade wharfhouse seen in 1949 and 1947 respectively. Almost exactly like the Cirencester building, this is better preserved and has for long remained a private house.

It is almost hidden from the modern road, and presents its large rear storage accommodation to the passer-by.

DUDGROVE DOUBLE LOCK

52. The best-known view of Dudgrove double-lock in action, dated 1814. A boat is in the upper chamber, with the lower chamber in the foreground. An impressive arrangement, Dudgrove may well have been less well-planned than is often thought. It lowered the level of the canal 11′ 6″ towards the Thames.

53 & 54. Inglesham round-house, lock and bridge and the entrance of the canal into the Thames. In the background, the spire of Lechlade church marks the nearest settlement on the river.

A superb grouping of three essential canal features, Inglesham is a fitting entrance into or exit from the Thames and Severn. 53. is early nineteenth-century, and 54. early twentieth.

55. Canal restoration is deservedly popular and the enthusiasm has spread to the Stroudwater and the Thames and Severn route. A Trust, recently formed from the old Stroudwater Canal Society, has already successfully completed its first stage between Bowbridge and Griffin's locks.

Here the dragline excavator works east of Stanton's Bridge in the background, and the scene harks back to the same area only a few years previously (no. 4).